BREEDING MINKS
IN LOUISIANA FOR THEIR FUR

LECTOR HOUSE PUBLIC DOMAIN WORKS

BREEDING MINKS IN LOUISIANA FOR THEIR FUR

WILLIAM ANDRÉ ELFER

ISBN: 978-93-93693-08-2

Published: -

© 2021
LECTOR HOUSE LLP

LECTOR HOUSE LLP
E-MAIL: lectorpublishing@gmail.com

BREEDING MINKS IN LOUISIANA
FOR THEIR FUR

A Profitable Industry

BY

WILLIAM ANDRÉ ELFER

PREFACE

This little volume is issued in illustration of the feasibility of breeding minks in Louisiana for their fur. It is the result of experiments conducted by the author himself, and he feels that it should be of interest to many and of value to the few who are looking for fields for profitable investment. It is the author's aim to issue a more elaborate work on the same subject sometime during the early part of next year.

W. A. E.

A Louisiana Mink. Notice the Small Eyes, and the Low, Rounded Ears, Scarcely Projecting Beyond the Adjacent Fur.

or the following description of the American mink I am indebted to the Encyclopædia Britannica:

"In size it much resembles the English polecat—the length of the head and body being usually from fifteen to eighteen inches; that of the tail to the end of the hair about nine inches. The female is considerably smaller than the male. The tail is bushy, but tapering at the end. The ears are small, low, rounded, and scarcely project beyond the adjacent fur. The pelage consists of a dense, soft, matted under-fur, mixed with long, stiff, lustrous hairs on all parts of the body and tail. The gloss is greatest on the upper parts; on the tail the bristly hairs predominate. Northern specimens have the finest and most glistening pelage; in those from the southern regions there is less difference between the under- and over-fur, and the whole pelage is coarser and harsher. In color, different specimens present a considerable range of variation, but the animal is ordinarily of a rich, dark brown, scarcely or not paler below than on the general upper parts; but the back is usually the darkest, and the tail is nearly black. The under jaw, from the chin about as far back as the angle of the mouth, is generally white. In the European mink the upper lip is also white, but, as this occasionally occurs in American specimens, it fails as an absolutely distinguishing character. Besides the white on the chin, there are often other irregular white patches on the under parts of the body. In very rare instances the tail is tipped with white. The fur, like that of most of the animals of the group to which it belongs, is an important article of commerce."

The fur market has always been a good market. It has grown firmer and stronger from year to year, while the prices for furs have been advancing steadily and rapidly with the growing demand for furs in Europe and America, and with the general increasing scarcity of all fur-bearing animals. Mink fur advanced about fifty per cent. during the last two seasons, and there is every reason to believe that the mink fur in Louisiana will advance to about six dollars within the coming three years. The minks caught in Louisiana last season were sold at an average price of three dollars.

Resting in a Warm Place. Notice the Long Body and Its Shape.

In a Position to Jump. Notice the Long Tail.

ur-bearing animals are becoming scarce where they were once so plentiful, and, like the buffaloes that roamed this country in such great numbers, they will soon, many of them, become extinct if the present rate of trapping continues to obtain in America. Already certain fur animals are almost trapped out and are rare. Even the alligator, which was so plentiful a few years ago in the swamps of Louisiana, is hardly sought after any more for its hide because of its scarcity.

The laws enacted by the various State legislatures for the protection of fur-bearing animals, in fact, offer no protection; for most furs caught out of season have no market value, and for that reason are not caught.

In Louisiana a trapper has to procure a hunting license if he wishes to carry a gun while trapping, which license costs only one dollar and is good for one season only. Such a low license, while it may bring a large revenue to the State, clearly has no element of protection in it. On the contrary, it is a truth that it stimulates both hunting and trapping, as there were more trappers in Louisiana last season than before the law requiring this license came into effect. Every trapper procures a hunting license whether he carries a gun or not, and most trappers believe the law requires them to have this license to trap.

Whatever is being done for the protection of fur-bearing animals in Louisiana, the fact remains that they are fast disappearing. Old and experienced trappers will tell you that minks were very difficult to trap last season as compared with the seasons of a few years ago, when they could be so easily trapped in dead-falls. Raccoons, too, which were so numerous in the rear of old cornfields during the trapping seasons, have diminished at a surprising rate within the last three years.

While laws are being adopted by different States for the regulation of trapping to protect fur-bearing animals, it is time for those who expect to make money with fur in the future to begin raising their own animals. The time is almost here when trapping will be unprofitable. Fur animals will be too scarce to make anything at it. Then people will have to build farms in which to breed minks for their fur, and mink farms will become common. Minks are the most valuable fur-bearing animals in Louisiana, being the most numerous, and they are also the easiest and most profitable to breed for their fur.

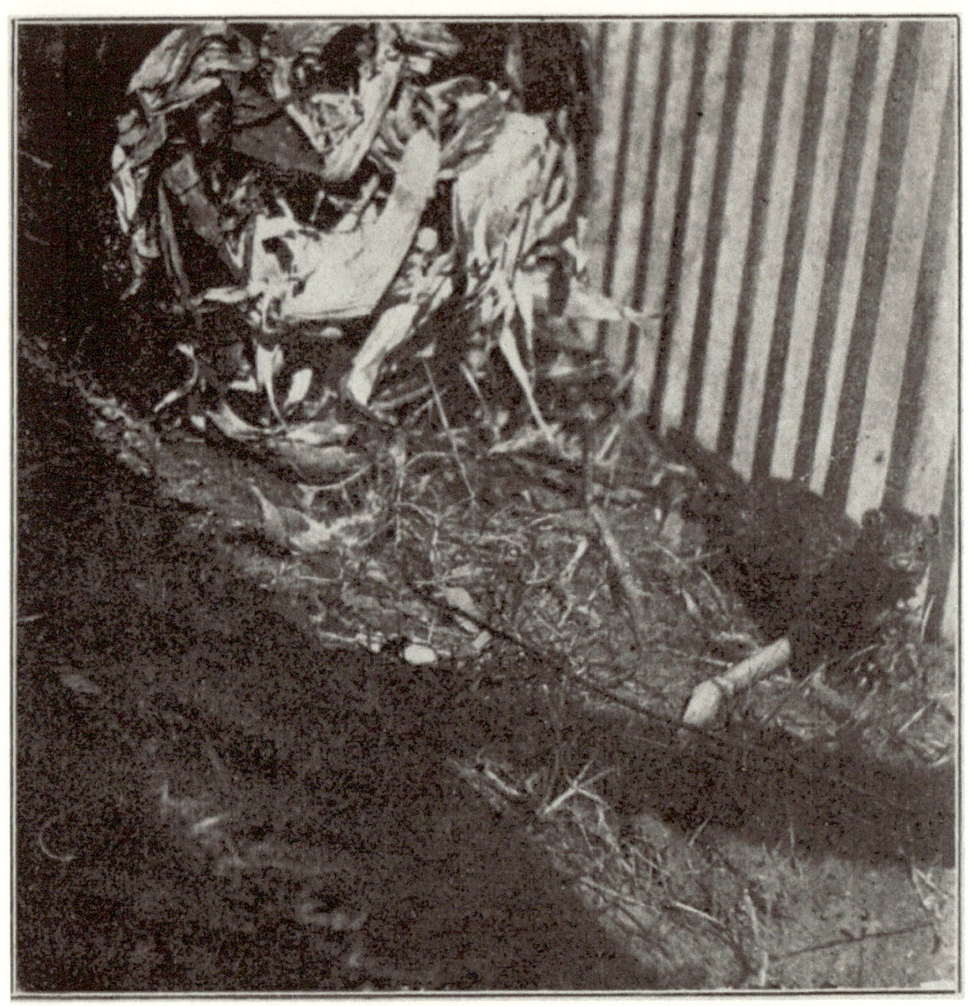

A Female of Two Years.

Breeding minks in Louisiana for their fur can be made a very profitable industry. There is more to be made at it than raising horses, hogs or cattle. After a farm is once completed and stocked, all expense is about over if there is a large-enough pond in it to supply the minks with sufficient food. Under the present condition of the fur market, each female will average a profit of forty dollars a year. A farm stocked for the first time during the winter with five hundred female minks should bring its owner the following winter approximately twenty thousand dollars. This is figured at three dollars a fur; but within three years the mink fur in Louisiana should be selling for what the mink fur in the North sold last season. With this increase in the price of fur, a farm stocked with the same number should bring forty thousand dollars.

Minks require little room, and thousands can be raised each year on a farm of ten acres. The larger the farm, however, the better chances they will have to procure food for themselves, as birds will enter a large farm more freely than a small one.

The Fur During the Summer Is Very Poor, and Not So Dark as It Is During the Winter.

For this reason, in building a mink farm the first and most important requirement is a good location. A small island consisting of low land covered with trees and grasses, with the opposite shore at least three-quarters of a mile distant, would make an excellent farm, provided the surrounding water supplies an abundance of small fishes. Such an island would, of course, preclude the necessity of using material for holding the minks in captivity. If a suitable island cannot be found, a good farm can be made with five or more acres of low swampy land having a natural growth of trees, grasses and underbrush, such as can be found in Southern Louisiana. But the piece of land selected for a farm must inclose a large pond, or several small ponds, containing a good quantity of small fishes, especially crayfish. The trees and grasses will attract birds, which, in addition to fish and rabbits, form a large part of food for the minks.

Feeding minks is pretty costly, and is hardly to be considered by one entering the business of breeding them for their fur.

An Excited Mink Trying to Climb.

The walls surrounding a mink farm can be made either with bricks or with sheets of corrugated, galvanized iron. The latter material makes an excellent wall, and costs less than a brick wall. It should be used in sheets measuring twelve feet in length by about twenty-six inches in width. These sheets should be used in an upright position, and at least five feet should be underground and seven feet aboveground. They should be allowed to lap two inches, and the dirt should be firmly packed against them. Two rows of wooden strips nailed on the outside of the wall, one about two feet above the ground, and the other along the top edge of the sheets, will greatly strengthen the wall and also prevent the wind from shaking it.

A Young Female Mink Walking Along the Walls of a Small Farm.

The following photograph shows a small pentagonal farm, the walls of which are made with sheets of corrugated, galvanized iron. Each side measures sixteen feet in length, extending four feet underground and four feet aboveground. Wire netting is used to cover the farm, not to prevent the minks from jumping over, although the walls are too low, but to prevent chickens, cats and buzzards from entering and eating the food put in for the minks. A wooden shed also covers a part of the small farm and serves to keep out some of the rain and heat, there being no shrubs or trees therein. There are two small troughs in the ground for holding water, and in the center of the farm there is a place for the minks to live during the day, which consists of boards laid five inches above the surface of the ground with about fourteen inches of dirt on top. Under these boards it is dark during the day and always damp and cool. There are also several barrels in this farm filled with corn shucks and hay for the minks to enter during cold weather. The minks in this little farm are fed with the spleen of cattle, different meats, crayfish and other small fishes. The cost of this farm, or pen, which has been used for experimental purposes only, amounts to approximately forty-two dollars. It is large enough to raise two hundred minks if they are properly fed and cared for.

A Small Mink Farm.

Part of Interior of Small Farm, Showing Boards With Dirt on Top for the Minks to Live Under During the Day.

Sometimes an island can be used for a farm even when it has opposite shores or islands within two hundred feet or less, provided the water surrounding it has an average depth of from four to six feet. In such a case, the walls inclosing the island should be built in the water at a distance of fifty or one hundred feet from its shores. Sheets of metal should be used, as previously described, by placing them upright in the water and nailing them together with strips running along the outside. It is not essential that the lower wall should be in the ground or even touching it; posts can be driven in the ground to strengthen the wall, or to support it entirely.

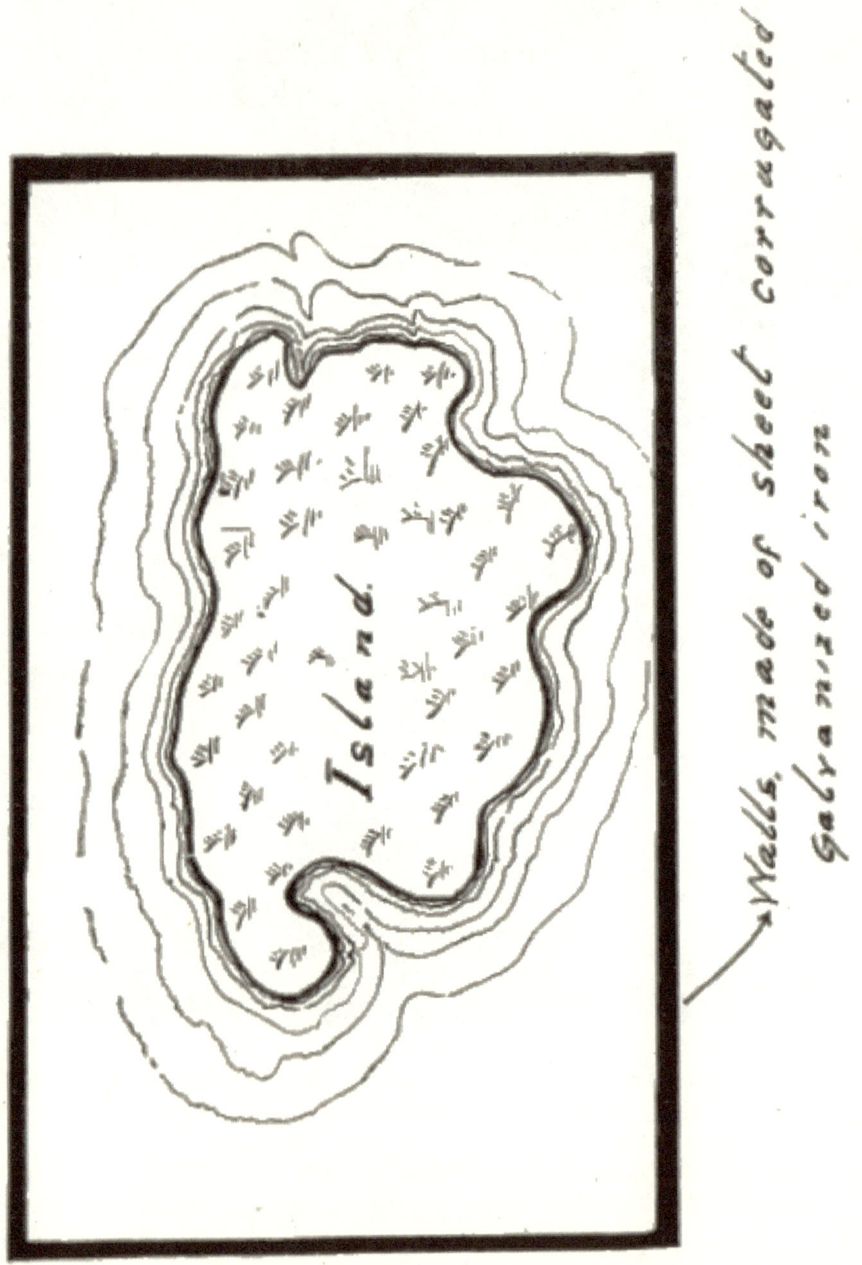

A Mink Farm Made Out of an Island. The Water Surrounding Has a Uniform Depth of Five Feet.

In a small farm where minks are in close captivity and have to be fed, the old ones used for the purpose of stocking it will at first do considerable digging near the walls. They will dig into loose earth to a depth ranging from a few inches to three feet in their attempts to liberate themselves. But they will cease to dig after they have been in captivity for about four months. Those born in a farm will not dig or try to get out. They will climb, however, to a height of fifteen feet on reclining trees or on bushes, and for this reason all trees, bushes and pieces of lumber should be removed from the inside of the walls before any minks are turned loose in a farm. They will ordinarily jump to a height of four feet. They can climb wooden walls as swiftly as a cat, or any wall made of soft material.

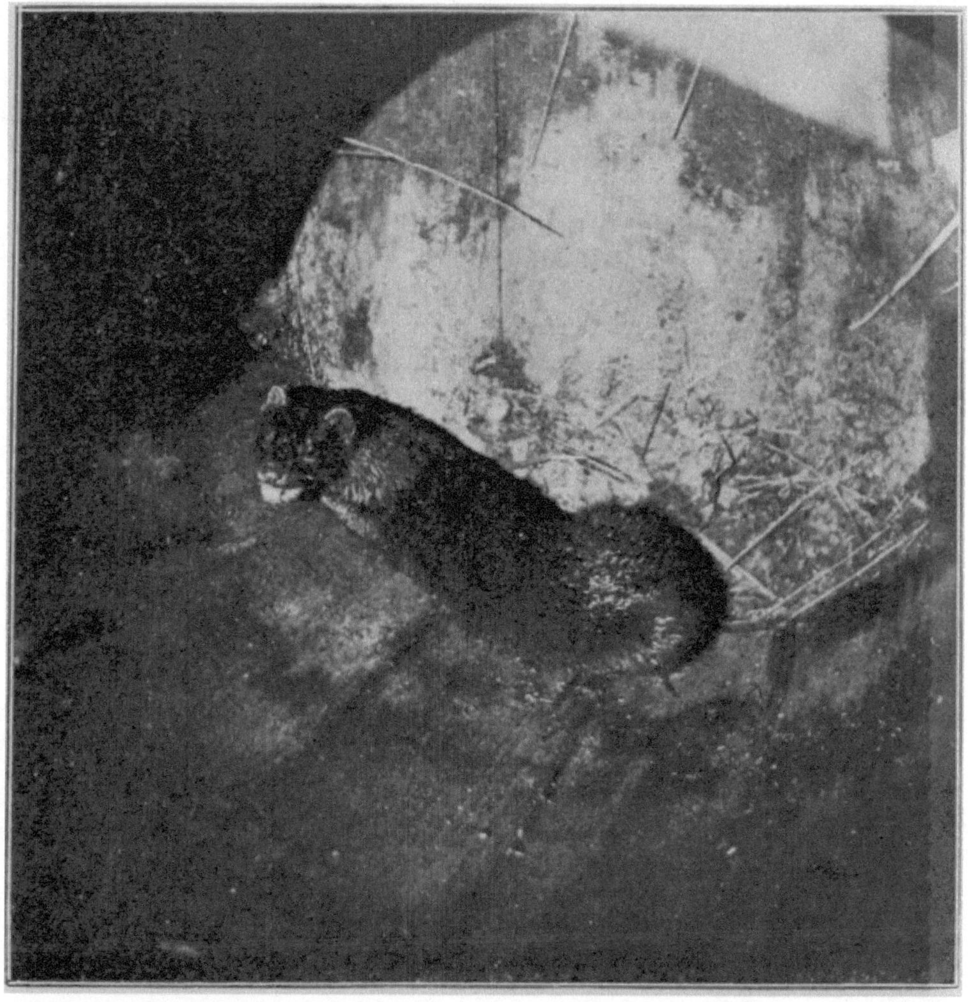

Disturbed in Her Sleep. Notice the Bushy Tail.

The following sketch shows the very best mink farm that can be made. It requires a rectangular piece of land of five or ten acres, running along and separated by a large bayou in the swamps of Louisiana. Covering this land there should be the necessary trees, shrubbery and grasses. The walls are built along the bayou about one hundred feet from the middle, and extend underground to a depth of six feet. The walls at the ends of the farm where they cross the bayou should be very carefully constructed. At these places where the walls cross the bayou should have a depth of at least twelve feet or more, so that the walls can be made to extend nine feet below the water surface for one-third the width of the stream and still have sufficient openings below the walls to permit the water to flow through freely. For example, if the bayou is fifty feet wide, fifteen feet of the wall crossing it can be elevated so that there will be a large-enough opening below for the water to flow. The remaining portion of the wall (that lying near the shore) should be driven in the ground for about one foot, as minks will not dig under water. A farm of five acres, similar to the one just described, would cost, completed, approximately eight hundred dollars. The minks in such a farm, owing to the continuous change of water in the bayou, would always have an abundance of food. The banks of the bayou would afford a natural breeding-place, as minks usually burrow in the banks of small streams or along canals and have their young near the water. If the water in the bayou falls, wire netting could be used over the opening at the ends below the walls.

> "Minks eat birds, small mammals and eggs. The principal food of minks comes from water, fish, frogs, crayfish." — *International Encyclopædia.*

The minks I have been experimenting with have persistently refused to eat frogs. I penned one up separately and attempted to feed her on frogs only, and I believe she would have starved rather than eat frogs.

Minks can be raised in any kind of pen or cage, and water is not essential to their happiness. They are easily tamed and like to be petted.

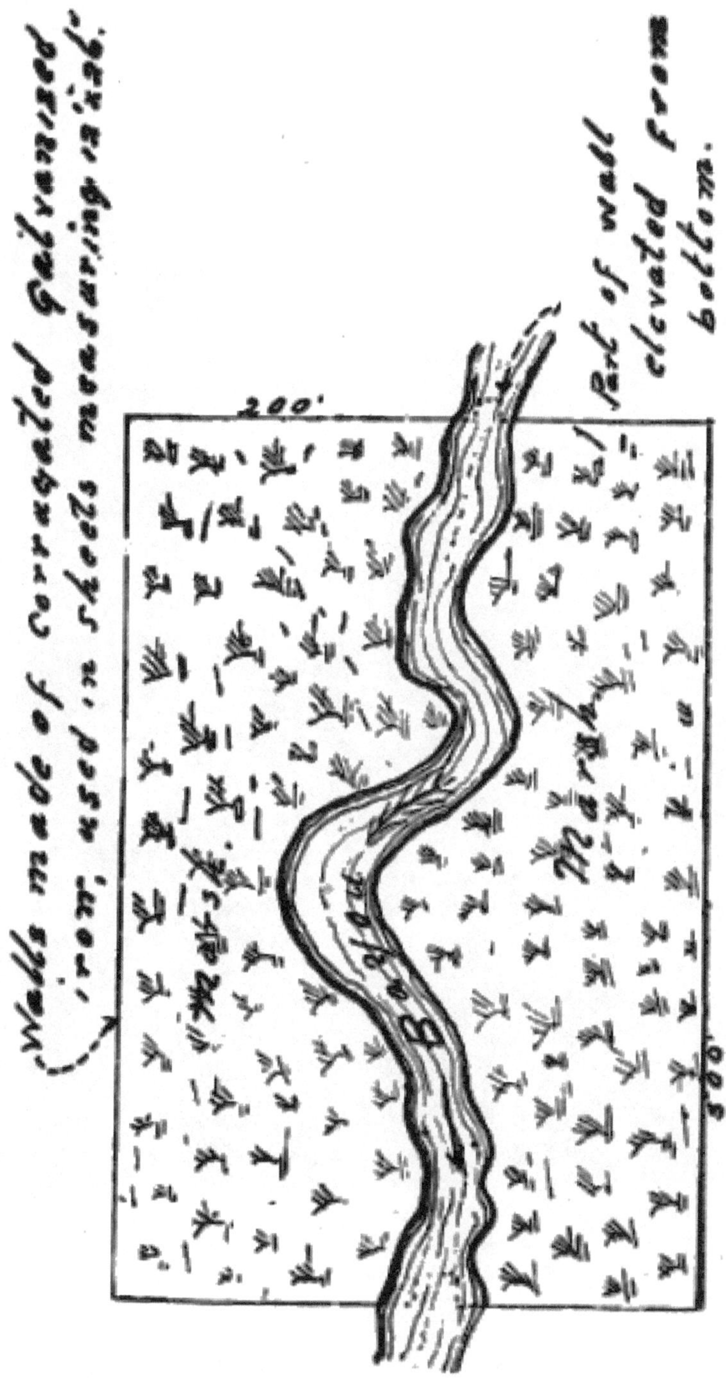

A Mink Farm Inclosing Portion of a Bayou, Allowing
the Water to Flow Through.

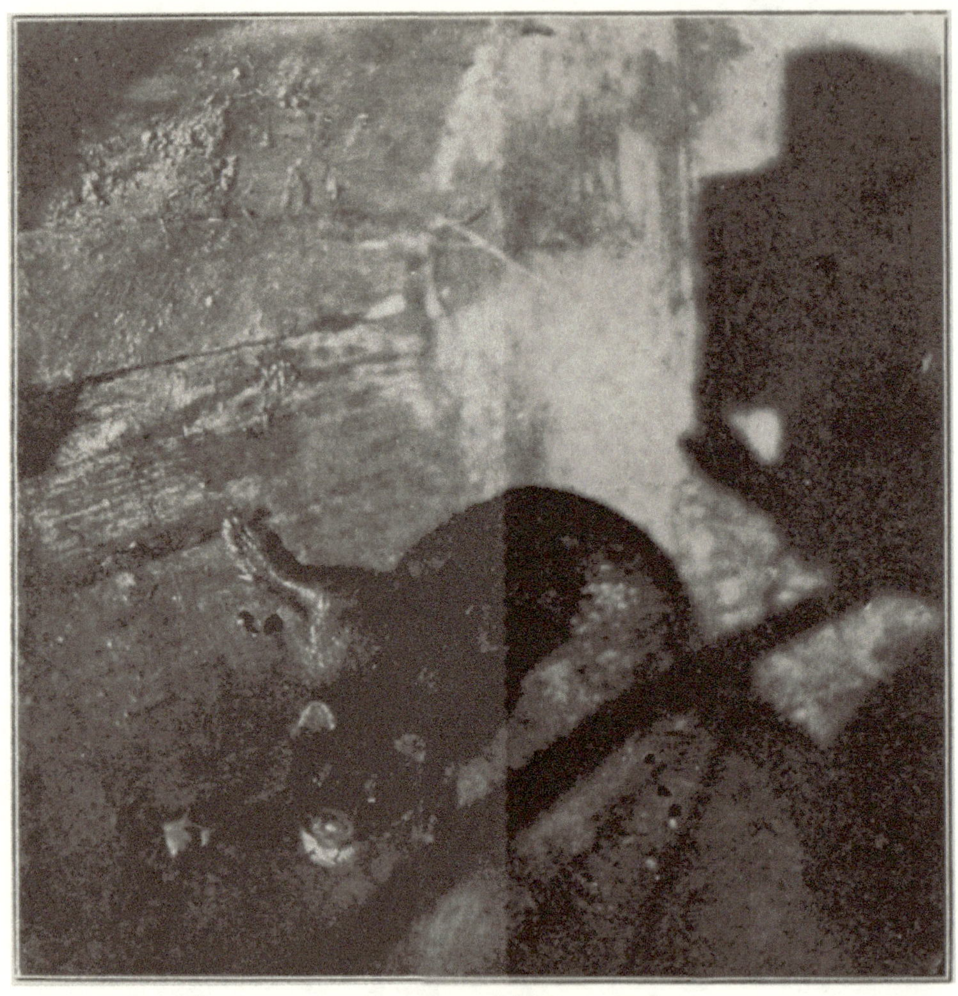

An Angry Mink.

HABITS OF THE MINK IN LOUISIANA

inks in Louisiana have two litters a season, the number of young in each brood varying from four to eight. Sometimes, however, but very rarely, there will be only two in a brood, and almost as infrequently, on the other hand, there will be three litters a season instead of two. Captive animals breed more profusely than the wild, and will occasionally have three litters where they are in close captivity. They begin to breed when they are about one year old, and in captivity will raise an average of fourteen a year. Normally, they live to be about nine years old, but they will live longer in captivity where they are well treated and given all the water and the different foods required by them.

Like all other industries, the business of breeding minks for their fur necessitates an outlay of capital. A farm cannot be built without money, and the cost of one sufficiently large to breed minks profitably ranges from five hundred to a thousand dollars. Of course, a farm can be made any size and costing any amount of money; but large farms are not necessary, and it is much better to have several small farms of six or ten acres than one very large one.

After a farm is completed it has to be stocked, and the task is no easy or inexpensive one. Trappers will have to be employed to trap minks with No. 1 steel traps, as these small traps do not injure them very much unless they are permitted to remain caught too long. Those that have badly-broken bones should not be bought, as suffering will cause them to eat their leg off, in which case they will always die.

The author intends to organize a company styled the "Louisiana Mink Company," the objects and purposes of which shall be to build mink farms and to breed minks in this State for their fur.

No matter what capital is involved, or expense incurred, in entering into the business of breeding minks for their fur, the returns will be so big that this will appear small in comparison. And those who are so fortunate as to start in the industry now will, when minks will have become so rare that trapping will be unprofitable, and the demand so great that the prices for mink fur will soar higher and higher—those persons, I say, of foresight, who had the good fortune to start in the business early, will reap each year the steady advances in the price of mink fur, and be able, in a word, to command the fur market of both Europe and America.

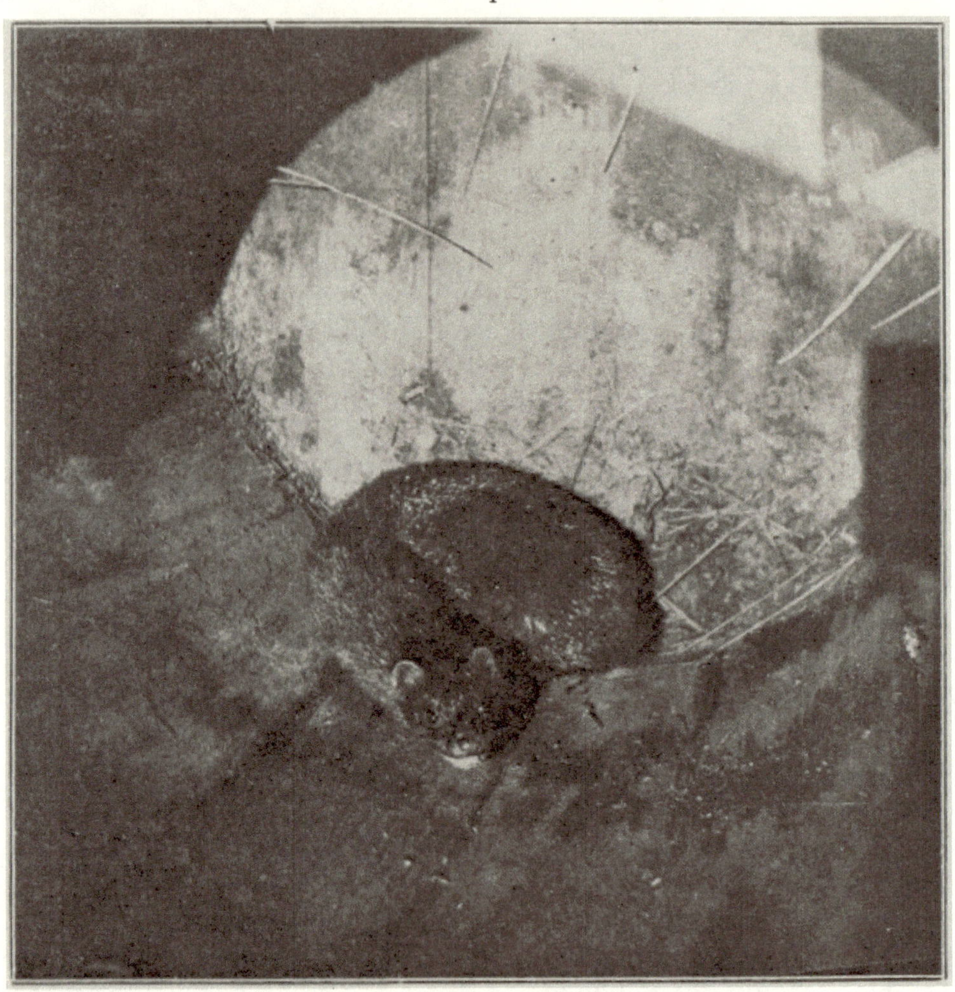

A Female Mink Resting With Eyes Open.

Lector House believes that a society develops through a two-fold approach of continuous learning and adaptation, which is derived from the study of classic literary works spread across the historic timeline of literature records. Therefore, we aim at reviving, repairing and redeveloping all those inaccessible or damaged but historically as well as culturally important literature across subjects so that the future generations may have an opportunity to study and learn from past works to embark upon a journey of creating a better future.

This book is a result of an effort made by Lector House towards making a contribution to the preservation and repair of original ancient works which might hold historical significance to the approach of continuous learning across subjects.

HAPPY READING & LEARNING!

LECTOR HOUSE

LECTOR HOUSE LLP
E-MAIL: lectorpublishing@gmail.com

www.ingramcontent.com/pod-product-compliance
Lightning Source LLC
Chambersburg PA
CBHW030424160726
47992CB00007B/3278